MathStart®
MEASURING

SUPER SAND CASTLE SATURDAY

by Stuart J. Murphy   illustrated by Julia Gorton

HarperCollinsPublishers

LEVEL 2

To Phoebe Yeh —
for her super support every
day of the week
　　　　—S J M

To Ivy, Raleigh, and Russell —
Who give me the widest smile,
who help me stand the tallest,
and for whom I feel the
deepest love　　　—J.G.

The illustrations in this book were done using airbrushed acrylic on Strathmore paper.

HarperCollins®, ■®, and MathStart® are registered trademarks of HarperCollins Publishers. For more information about the MathStart series, write to HarperCollins Children's Books, 10 East 53rd Street, New York, NY 10022.

Bugs incorporated in the MathStart series design were painted by Jon Buller.

SUPER SAND CASTLE SATURDAY
Text copyright © 1999 by Stuart J. Murphy
Illustrations copyright © 1999 by Julia Gorton
Printed in the U.S.A. All rights reserved. http://www.harperchildrens.com

Library of Congress Cataloging-in-Publication Data
Murphy, Stuart J., date
　Super sand castle Saturday / by Stuart J. Murphy ; illustrated by Julia Gorton.
　　p.　cm. — (MathStart)
　"Level 2, Measuring."
　Summary: Introduces the concept of nonstandard measurement as three friends compete in a sand castle building contest.
　ISBN 0-06-027612-6. — ISBN 0-06-027613-4 (lib. bdg.) — ISBN 0-06-446720-1 (pbk.)
　1. Mensuration—Juvenile literature. [1. Measurement.] I. Gorton, Julia, ill. II. Title.
III. Series.
QA465.M87　1999　　　　　　　　　　　98-3210
530.8—dc21　　　　　　　　　　　　　　CIP
　　　　　　　　　　　　　　　　　　　AC

Typography by Julia Gorton
　　4　5　6　7　8　9　10

It was a hot and sunny Saturday— a super day for a sand castle contest.

"Okay, everybody," shouted Larry the Lifeguard. "You need to finish building your castles before the tide comes in."

"Today I'm giving out prizes for the tallest tower, the deepest moat, and the longest wall.
Ready? Go!" Larry blew his whistle.

Juan, Sarah, and Laura
couldn't wait to get started.
They filled their pails
and got right to work.

Sarah wanted to win the prize
for the tallest tower.
She started building
up, up, up.
But Juan's tower
was getting tall too.

"I'll bet my tower is taller than yours," said Juan. "Let's use our shovels to measure them."

"Great idea,"
agreed Sarah.

"My tower is three shovels tall!"
said Sarah.

Then Juan tried to dig the deepest moat. Nearby, Laura was digging a deep moat too.

"Let's see whose moat is deeper," said Laura. "Our shovels are too big, so let's use our spoons to measure."

"Okay," said Juan.

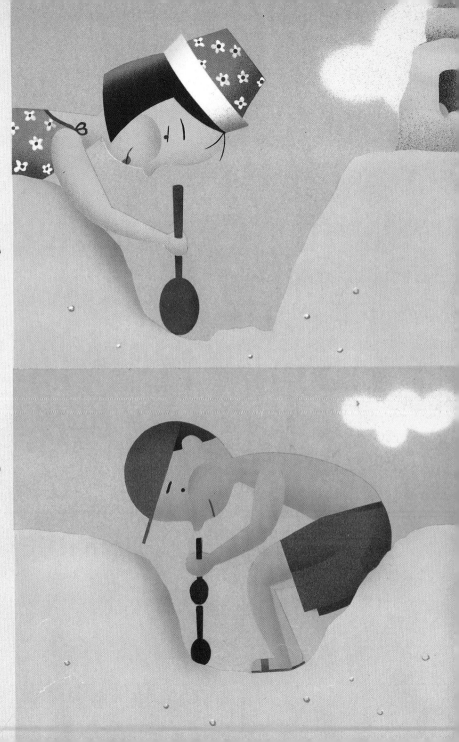

Juan shouted,
"My moat is
two spoons deep!"

"Mine is a little more than one spoon deep," moaned Laura.

"Hurry up, you guys!"
Larry called.
"The tide's coming in fast!"

"Laura!" called Sarah. "I just finished my wall.

Let's see whose is longest — it will be easy to measure if we use our feet!"

They quickly
walked
heel to toe
along their
walls.

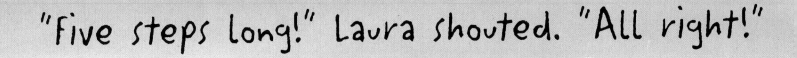

"Five steps long!" Laura shouted. "All right!"

"Sorry," said Sarah. "Mine is seven steps long. I win."

Larry blew his whistle again.

"All right, everyone!" he shouted.

"Time to see who won."

Larry measured
and measured,
and then he
measured some more.

"Juan has the TALLEST tower," announced Larry.

44 inches

48 inches

40 inches

Sarah's tower

Juan's tower

Laura's tower

"But why, Larry?" Sarah asked.
"My tower was three shovels high, and Juan's was only two."

"Your tower was more shovels high, Sarah," Larry explained,
"but your shovel is shorter than Juan's."

"Laura has the DEEPEST moat," said Larry.

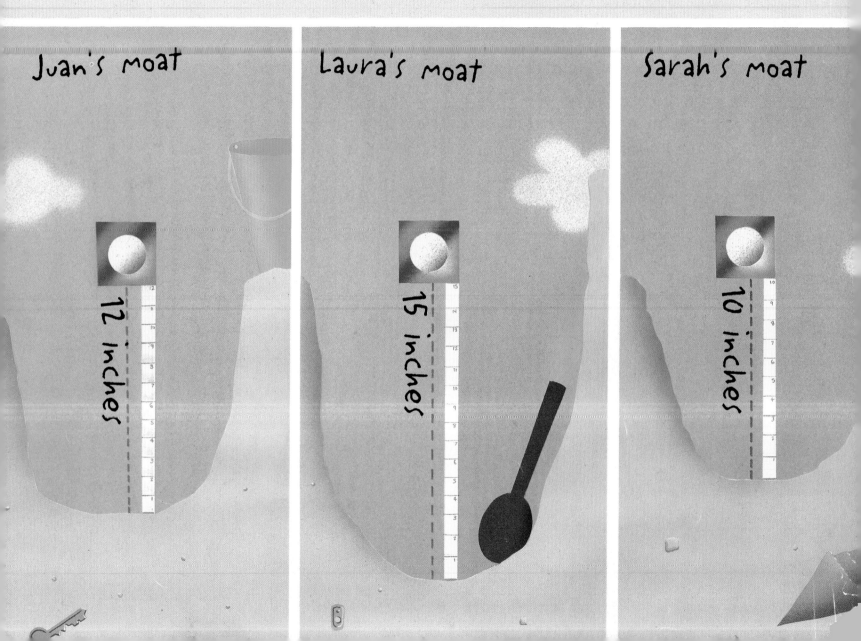

"Juan's moat measured more spoons deep than Laura's, but his spoon is shorter than Laura's."

"And Laura has longer feet,

but her five-step wall still didn't beat Sarah's seven-step one.

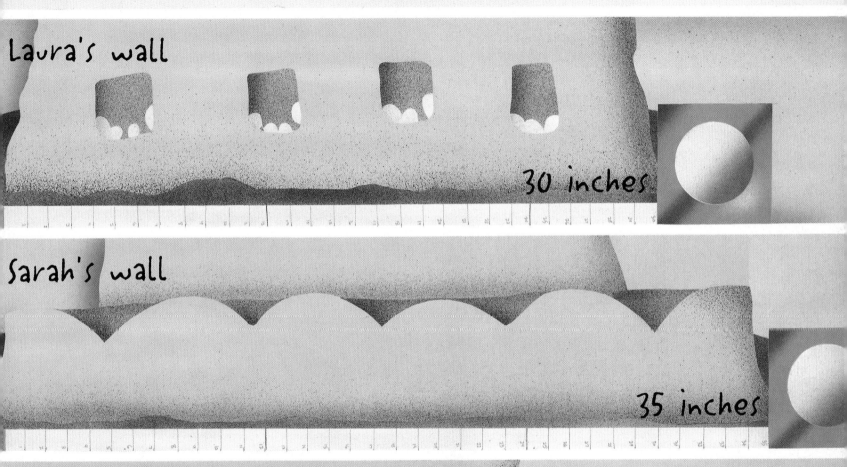

Laura's wall

30 inches

Sarah's wall

35 inches

Juan's wall

24 inches

"Spoons and shovels
and people's feet
can all be different sizes,"
said Larry,
"but an inch
is always an inch."

If you would like to have more fun with the math concepts presented in *Super Sand Castle Saturday*, here are a few suggestions:

- Read the story with the child and talk about what is going on in each picture. Discuss why measuring the castles is a good way to check to see who has the tallest tower, the deepest moat, and the longest wall.

- Talk about the pages that show the measuring taking place. Ask the child to describe what is going on in each case. Point out that the measuring must be end to end for it to be accurate.

- Ask questions throughout the story, such as: "Do you think that using a shovel would be a good way to measure the tower of a castle?" and "Is a spoon a good way to measure the depth of a moat?" Explain that these tools can be used for measuring, but that tools of the same length must be used consistently.

- At the end of the story, ask why it is sometimes helpful to have a measuring system that everyone uses.

- Pick distances around the house and measure them using "baby steps" and "giant steps." Is the hallway more baby steps or giant steps long? Are there more baby steps or giant steps between the couch and the TV? Explain.

Following are some activities that will help you extend the concepts presented in *Super Sand Castle Saturday* into a child's everyday life.

Body Lengths: Have friends take turns lying down on the floor and measuring each other from head to toe using straws, and then a ruler. Make a chart that shows the length of each person in terms of each unit of measurement.

Estimation: After having had some experience with linear measurement, try estimating the lengths of household objects in inches or centimeters. Check the estimates with a ruler.

Moving Furniture: Measure the width and height of a bookcase or another large piece of furniture. Then determine other places where it could fit.

The following books include some of the same concepts that are presented in *Super Sand Castle Saturday*:

- TWELVE SNAILS TO ONE LIZARD: A TALE OF MISCHIEF AND MEASUREMENT by Susan Hightower

- MEASURING PENNY by Loreen Leedy

- INCH BY INCH by Leo Lionni